Changing Your Mind

A Theory of Space without Time
t=cB

I0471829

MARC E. KING

To order additional copies of this book, contact:
Xlibris Corporation
1-888-795-4274
www.Xlibris.com
Orders@Xlibris.com

Forward by the Author

This book can by understood and enjoyed by everyone.

The text is somewhat technical; however, it must be technical in order to sustain the criticism.

The mathematical model is sound. Once a model can be defended mathematically and physically, then much of the criticism becomes superficial.

Some very bright physicists have already disputed the model. Unfortunately, when asked about their own views, their answers have been less than enlightening.

For example, what is time? "Well, it seems to be continuous, never ending, directional, and a thing that describes other things… everyone knows that." The answer is not acceptable.

This text explores the intuitive and mathematical relevance of the so-called real-and-continuous variable $t = $ time.

Required proofs are documented, but the mathematics is not terribly difficult and can arguably be understood at an advanced high school level. There is a non-mathematical supplement that should be more intuitive and pictorial. The mathematics is not so important to understand relative to the basic dimensional concept.

The difficult concepts are dimensional and are hard to imagine in our three-dimensional mindsets. The reader should assimilate and internalize the text. With "time," the concepts should become clearer from day-to-day experience after reading the text.

Time is a valuable concept. It measures day and night as well as seasons of the year among many other realities. No one should throw away their clocks upon reading the text! Macroscopically, this model is exactly the same as classical physics.

Please feel free to ignore all appendices in the technical and non-technical segments. The appendices more rigorously define and prove the text but are not required to understand the concepts.

Regarding the mathematics, the author has taken the usual and expected physicists' liberty with rigor.

Contrary to some criticism, this text in fact fully supports the concept of Deity.

Table of Contents:

I. Introduction

We have evolved over hundreds of millions of years on the surface of the planet Earth. Our evolution has taught us nothing about anywhere else. Our perceptions are ingrained in us through our experience on the planet surface.

Most of us have never been anywhere else except here. The few who have been in orbit about the planet or in deeper space have been so close to the planet surface that their experience has been largely unchanged and innocuous in a relative way. Our surface-ingrained perceptions, our inherited and assimilated perceptions, the perceptions that were in fact created within our physiologies through evolution on the planet surface are very difficult to change in our minds.

Everyone wants to better understand our physical universe. This book and manuscript suggests we cannot achieve a higher knowledge while using our confined, local, and captive understanding and perception. The universe is not required to conform to our local perception. The universe is not a part of our mathematics. In fact, our mathematics is one small part of the universe and is only valid in and through our present and local perceptions.

In order for clear thought in an open-minded environment, we need to step aside from our captive perceptions. This writing suggests that our minds must and/or will change in three ways:

First, our concept of time is a perception constructed through and based upon hundreds of millions of years of evolution in a single place, i.e. the Earth surface. Our ancestors did not live anywhere else. Our perceptions, our blood pressures under the Earth's force of gravity, our neurology, and everything about us has evolved here and only here. We need to re-evaluate our local concept of time. Not simply in the previous "general-relativistic" way, but in the actual concept of continuity and in the new science and new intuition of time per this modern text.

Related and secondly, we must not allow our thinking to be dictated, banned, or controlled by anyone for any reason.

As an example, many empirical facts from existing data from the plethora of geosynchronous orbiting satellites are confidential and proprietary. This text explains exactly why the satellite clocks do not know what time it is "up there" and require continuous error-correction.

Thirdly, readers will soon suspect that everyone's mind may operate differently from their previous perceptions.

The common perception of the brain is that it is something like a huge magic super parallel processor. That it comes equipped with infinite ROM (read-only-memory) and even a little communication ability as part of the bundle.

Nothing is farther from the truth.

To start, please think about any important memory, e.g. a high school prom, your first remembrance, a first date, a favorite teacher or professor, yesterday morning, a friend or relative, anything at all.

A photographic type image immediately appears to you for the memory you have addressed. There is no end to how many of these photo-type images you can visualize. Basically, an infinite number of images.

Do you truly believe there is a magical infinite ROM (read-only) memory storage within the size of a grapefruit sitting above your shoulders? Infinite? Never ending? The size of a grapefruit?

Most likely not.

We suggest a more realistic view. Besides a better understanding of hydrocarbons from a new model, this text also suggests the brain as described above is more of an "adequate" processor with no ROM (zero read-only-memory) but having an extremely capable communication ability.

In fact, the exact opposite of our old perceptions!

As asserted in the Forward, this book needs to begin with the solid technical justification proved in the manuscript. The mathematics can arguably be understood by a high school math student. The math is not so important as the concept. You do not need to follow the proofs line by line; instead, the reader can look at the results along with the statements of results. There is an illustrated non-technical section following the technical justification.

Are you ready to open your mind and challenge your evolutionary and local perceptions?

Within this text, we show that energy production and obtainment is far easier and cheaper than our present understanding allows. We show that longevity of hydrocarbon molecules-of-life can and should be much more robust than our existing perceptions allow.

In this text, we prove that time is a valuable and necessary measurement tool, but that it is not physically real; instead, reality is a sequence of spatial frames similar in some ways to a photographical sequence that has been aligned in order to make a movie (motion picture.)

There is no real continuous time t. Our perceptions have misled us.

The Transformation

II. Manuscript:

A Mathematical Transformation of Variables Defining Space – Time and the Constant h

Marc E. King

Silicon Valley, California

June, 2012

Abstract

A variable transformation for time t is supported by wave mechanics and relativity theory and shows that time and space can be related and connected by the concept of physical events per unit space. The transformation confirms our daily macroscopic experience as unchanged from classical physics while still suggests new physics regarding small energies and large spaces. Perceived time can be altered relative to Earth-bound clocks in regions of lower or higher gravitational force. A series of calculation-verifications proves the theory, derives Planck's constant and defines quantum mechanics. Black holes and their mass-radius relationship are defined. The Schwarzschild Radius is Defined. Minimum and maximum energies are defined.

Introduction

In this model, it is shown that continuous time t and a contiguous view of spatial frames are mathematically the same in the macroscopic sense. A suggested transformation of variables presents interesting differences in concept for small and large energies and spaces.

Concept

We postulate that continuous time as experienced can also be represented, with the same physical result, as a directional spatial sequence or frames of events.

We consider a new unit system using the transformation $t = cB$ with c = speed of light, where one spatial frame (size b) is related to one physical event B by

b (meter) = 1 (event) / B (events meter^-1)

The transformation $t = cB$ implies the units t (sec) = c (met/sec) x B (sec^2 met^-1)

Then B events per meter = B sec^2 per meter,

and one physical event = one square second = sec^2.

Derivation

From wave mechanics, we have the Schrödinger equation[i]

$d\psi/dt = +/- 2\pi i/h \times E\psi$ as a partial derivative

and the related approximation

$\Delta x \Delta k \geq O(1)$.

This defines the uncertainty in measurements [ii]

$\Delta x \, \Delta p \geq h/2\pi$.

Implying $\Delta p = m \, \Delta x / \Delta t$ and using a transformation for Δt, then

$\Delta p = m \, \Delta x / \Delta(cB)$

This leads to

$m (\Delta x)^2 \geq (h/2\pi \text{ J-s}) (\Delta cB) = (h/2\pi \text{ J}) (cB) (\Delta cB)$

Per unit mass, then

$(\Delta x)^2 \geq (h/2\pi)(\Delta cB)(cB)$ from transformation.

For a single B (events-meter^-1) the corresponding $\Delta x = b$ meters and $\Delta(cB) = 1/(cB)$

Then $b \geq$ (h-bar)^1/2.

Further defining b-minimum as the minimum Δx and using the positive root in this analysis, then

b(min) =1.027E-17 meters.

Subject to the further justification below, we assert:

E = F-sub-B x b

Where F-sub-B = F-sub-G = the gravitational force at the spatial location of event B.

And on the planet surface, F-sub-B = ma = m x 9.8 meters-sec^-2.

Then E / m = a x b = 9.8b meters / (cB)^2 or

E / m = 9.8b / (c/b)^2 and

E / m = 9.8 / 9 (10^-16) b^3 J-kg^-1,

Or we can write the expression:

E / m / b^3 = 1.089E-16 J kg^-1 or

E-sub-B / m = 680 eV / kg for one cubic spatial boundary.

Justification for Spatial Dimension b

Using uncertainty and similar to the derivation above, one estimation using neurological sensory communication as an upper bound on Δx = b (in one dimension) for spatial boundary (frame) size is suggested by:

(Spatial Frame Width)^2 ≤ (h – bar) x (c) x (Time Required for Sensory Continuity)

Using orders of magnitude 10^-34 J-sec (and adjusting for units) from wave mechanics and estimating the time required for sensory communication in the range 10^-3 sec – 10^-6 sec from synapse switching (potential change) rate, we would then estimate the magnitude:

Δx = b ~ 10^-14 to 10^-16 meters (for example as an upper spatial bound) in order to perceive continuity from actual contiguity of frames.

The neurological bound approximates the largest frame or spatial size that could be perceived as the continuity of time and accommodates the calculated boundary dimension Δx = b ~ 10^-17 meters.

b(min) = 1.027E-17 meters was derived assuming t = cB so that c is assumed to be the maximum achievable velocity[iii] and as such defines a maximum sequential rate and a minimum allowable b.

Justification for Spatial Barrier Energy

Using a one dimensional example,

E = F x distance.

We are using the transformation $t = cB$ where $B = 1 / b$ and b has the spatial dimension of meters.

The energy associated with the distance b is a function of a force F acting upon a mass m at a particular set of spatial coordinates.

It follows, the innate force acting on the mass m in space is the gravitational force.

There are no external forces to be considered for the mass m for our purpose regarding the transformation associating time and space.

Verification of Calculations

E-sub-B is then a function of gravitational force.

Using the calculations above and for a single event $B = 1$,
we can also write, for the planet surface as an example:

$E / m = a \times b = b \times 9.8 (c / b)^{-2} = 9.8 / c^2 / b.$

Then $1.089E-16 = 9.8 / c^2 / b.$

And for the planet surface E-sub-B, we verify our unit of measure calculations:

b meters = (1 event / B events meter^{-1}) = 1.000 as a confirmation of the energy calculation 680 eV / kg.

Independently, we can re-calculate the value of b using F-sub-G on the planet surface:

b = E-sub-B / m x (a)$^{-1}$ = 1.089E-16 / 9.8 = 1.111E-17 meters.

This should be the universal value of b and is independent of F-sub-G since the accelerations "g" cancel for any spatial position.

This value is larger than the allowed minimum calculated b(min) = 1.027E-17 by the difference 8.4E-19 meters and we find the calculated surface value to be approximately 8% larger than the minimum allowed value b(min) using the Earth gravitational acceleration a = g = 9.8 m-s^{-2} and using no transformations in this calculation. We do not pursue further calculations in the present scope. **(See Appendices for calculations.)**

Energy Change as a Function of F-sub-G = F-sub-B

Assuming mass m and boundary b are unchanged, then E-sub-B changes as a function of F-sub-G = F-sub-B the gravitational force at the location of physical event B.

This follows directly from

E-sub-B = F x b.

A smaller gravitational force leads to a smaller E-sub-B relative to the planet surface.

With different E-sub-B, clocks should appear to run at different rates in regions of higher or lower F-sub-G relative to the planet surface.

A fictitious force, like the Coriolis force or the weightlessness of orbit, should not affect the real force F-sub-G = F-sub-B.

Conclusions

Continuous time can be represented by a contiguous spatial sequence of frames, or boundaries $\Delta x = b$, while conforming to existing physics in the macroscopic sense and with our sensory perceptions.

The transformation t = cB leads to the spatial frame dimension b(min) = 1.027E-17 meters and corresponds to

3×10^8 physical events in one second of time t.

For this model,

The surface barrier energy E sub-B per unit mass = 680 eV / kg has been defined.

E-sub-B is suggested to be a function of gravitational force and so a function of spatial location.

Perceptions of Earth-time and clocks are expected to experience different rates in regions of lower or higher gravitational force relative to the planet surface.

For the following appendices A through C, please note: while the mathematics is a fact, the reader may question the physical relevance of appendices A through C if desired while independently understanding the physics of the following appendices D and E.

Appendix A
Fibonacci Calculations and Continuity

The difference in value between b(min) derived from the Schrödinger equation and b calculated empirically from Earth-surface F-sub-G is 8.4E-19 meters and proves to be the mean factor 1.079 or 7.9%.

The concept of continuous time t leads to exponential growth-decay:

$a = a\text{-sub-0} \times e^\wedge (\text{rate} \times \text{time})$ where $e = \lim (n \to \text{infinity}) (1 + 1/n)^\wedge n = 2.718$.

Applying the expression for time itself, then

$T(\text{new}) / T(\text{old}) = e^\wedge(r \times t) = e^\wedge(0) = 1$.

For time t itself, the rate $r = 0$ and there is no change in continuous time t so that one "second" of "time t" does not change. Time t is absolute.

Differential equations for centuries, e.g. the Schrödinger equation, assume time t is a real and continuous variable.

In the transformation $t = cB$, we need to treat continuity of time as a slight-contiguity of space.

In that case, we find $3 \times 10^\wedge 8$ met/sec to be a large enough frequency (number n) to continue using the calculated value of $e = 2.718 = \lim$ as $n \to$ infinity of $(1+1/n)^\wedge n$.

But in a directional spatial sequential model, then space itself advances or grows per some rate different from $r = 0$.

As a one dimensional chalk line curves in a two dimensional blackboard, and as a two-dimensional earth-surface curves in three-dimensional space, then a change in 3-dimensional space needs to take place in a mathematical dimension higher than 3.

Asserting the higher number of dimension (vertices) to be 5 as in the Fibonacci[iv] infinite sequence, and considering physical events $B = 1 / b = \sec^\wedge 2$, then we calculate the following for one second of time t:

$(V\text{-sub-S} / V\text{-sub-0})^\wedge 1/5 = e^\wedge(r_v \times t)^\wedge 1/5 = e^\wedge(.618^\wedge(5-3))^\wedge 1/5 = 1.079$

or a 7.9% decrease in physical events B (increase in one-dimensional size b) from the continuous-time model used to calculate b(min).

Appendix B

Another Calculation for difference b-empirical - b(min) = 7.9%:

In two dimensional physics,

$F = ma = m \times met\text{-}sec^{-2}$ becomes $F\text{-}sub\text{-}2 = m \times a\text{-}sub\text{-}2 = m \times met -sec^{-3/2}$

and one physical event B would no longer have units of sec^2; instead, $sec^{3/2}$.

The uncertainty principle then has a transformed h-bar, and now

$b(min) = (h\text{-}bar)^{1/2} \times c^{1/2} = 1.779E\text{-}13$ meters.

Similarly, $E\text{-}sub\text{-}B / m = 9.8 / c^{3/2}$ J-kg^{-1} per square boundary = 1.886E-12 J-kg^{-1} per boundary and

$b = E\text{-}sub\text{-}B / m / 9.8 = 1.924E\text{-}13$ meters.

Then we again have the mean factor

$= 1.079$ or 7.9%

between b(min) and b (from F-sub-G) in two-dimensional space exactly the same as in three-dimensional space.

Appendix C

A General Fibonacci Calculation:

The Fibonacci infinite sequence was referenced in Appendix A,

$F(n) = F(n-1) + F(n-2)$ with seed values $F(0) = 0$ and $F(1) = 1$.

Ratios converge, and

$\lim(n \to \infty) F(n+1) / F(n) = \varphi = (1 + 5^{1/2}) / 2 = .618 \ldots$ and

$\lim(n \to \infty) F(n-2) / F(n) = \gamma = .382 \ldots$ and so on.

Writing an example expression for spatial dimension ≥ 3 per Appendix A

$$\iiiiiiint dV = \iiiiint dV(0) \times \exp(r_V \times t)$$

where $r_V = r\text{-sub-}V = \varphi^{\wedge}(D(n+1) - D(n))$

then

$dx / dx(0) = \exp(\varphi \wedge (D(n+1) - D(n))^{\wedge}1/(n+1)$.

Except we are now doing math in another dimension, and while $e = 2.718$ in three dimensions, the base of natural logarithms should change in higher or lower dimensional space.

For example, in the case of 8 dimensions: $e \to e^{\wedge}(1/\gamma)^{\wedge}5/2$.

We quickly find $dx / dx(0) = 1.08$.

A different example, for the case of spatial dimension < 3:

The base e must change as a function of the power of B, i.e. in three dimensional space $B \sim \sec^{\wedge}2$ while in two-dimensional space $B \sim \sec^{\wedge}3/2$.

The difference in power of physical events B

$2 - 3/2 = 1/2$ and the two-dimensional $e = 2.718^{\wedge}1/2$.

Then we quickly find $dx / dx(0) = 1.08$ similar to the previous mean calculations for the difference between b(min) and b-empirical.

The (Fibonacci) calculations hold true for any spatial dimension n moving through n+1 with a dimensional adjustment for e.

Quantum Mechanics for t=cB

III. Quantum Mechanics without Time

For the previous appendices A through C, please note: while the mathematics is a fact, the reader may question the physical relevance of appendices A through C if desired while independently understanding the physics of the following appendices D and E.

We prefer to use the bound energy states for the Hydrogen atom as one example of quantum mechanics. We will focus on the lowest two states and transitions shown below and we immediately notice the energies for n=1, n=2, and the transition from n=2 to n=1 readily sum to the exact quantum barrier energy E_B per the main text.

$$
\begin{array}{lll}
n=\infty & \text{-------------} & E = 0 \\
n=4 & \text{════════} & E = -0.85 \text{ eV} \\
n=3 & \text{————————} & E = -1.51 \text{ eV} \\
\\
n=2 & \text{————————} & E = -3.40 \text{ eV} \\
\end{array}
$$

$$
\begin{array}{lll}
n=1 & \text{————————} & E = -13.6 \text{ eV}
\end{array}
$$

Appendix D

Definition of Planck's Constant

Planck's constant[v] $h = 6.626068E-34$ met^2 kg sec^-1:

From the Schrödinger equation,

$h = 13.6eV / (1^2) / v = (13.6eV / (1^2) / c)$ x 91.2nm

Then $h = (E\text{-}sub\text{-}B / c)$ x (91.2E-9 / 50)

Or $h = (E\text{-}sub\text{-}B / c)$ x 1.82E-9

And $h = (E\text{-}sub\text{-}B / c)$ x (b x c / 1.82)

So

$h = E\text{-}sub\text{-}B$ x b / 1.82 or

$h = e^{(-3/5)}$ x b x E-sub-B = b E_B / $e^{3/5}$

where

b = 1.111E-17 meters

and

E_B = E-sub-B = Earth surface barrier energy = 680eV/kg = 1.089E-16 J/kg

And the calculated $h = 2.718^{-3/5}$ x 1.111E-17 x 1.089E-16 = 6.6E-34 per event.

More precisely from our 3-decimal calculations and per appendices A through C,

h $\rightarrow$ h(1-Δh) where Δh = 0.08^5/2 and h = 6.63E-34 per event or we can write

$h = b E_B \kappa = b E_B (1 - \Delta h) / e^{3/5} = b E_B (\gamma' / \rho')$.

Units for transformed h:

h ~ met J Kg-1 b-3 ~ met-2 J Kg-1 ~ met-2 met2 sec-2 Kg Kg-1 ~ (one B)^-1 = event-1.

Appendix E

The Nature of Quantum Mechanics

We assert that any allowed energy quanta has a wavelength $\lambda = nb$ where n is an integer and $b = 1.111\ldots$ E-17 meters per the main text.

For example, the 13.6eV H ground state transition is $\lambda = 91.2$nm

And $n = \lambda / b = 820882088$.

Similarly, the H state 2 to state 1 transition is $\lambda = 121.6$nm

And $n = 109450945$.

An H state 4 to state 2 transition is $\lambda = 486.1$nm and $n = 43753375$.

Then $0 <$ one energy-event $\leq c / b$ (= 2.700E25 J-event = c3 J-event)

and quantum energy = hc/λ is defined as an integral operation of $1/b$.

Then the base of all quantum mechanics is $1/b = B$ where $t = cB$

and $h = h(a_G)$ and becomes a function of $F_B = F_G$.

It seems better to write the equivalent expressions:

$E = hv = hc/\lambda$ leading to

$E\ (J) = (b\ E_B\ \kappa)$ event^-1 x $c / (nb)$ energy-event and

$E / E_B = (\kappa / n)\ c$ or

$n\ E / E_B = \kappa\ c$

21

where:

$1 \leq n \leq \kappa c$ for E greater than E_B, and

$\kappa c \leq n \leq n_{MAX}$ for E less than E_B.

And this more clearly defines the universal nature of quantum mechanics.

Units for the ratio E / E_B are mass-spatial boundary, and the free-energy-state becomes E_B (per unit mass) instead of "zero."

Beyond the outline of this appendix, it can be shown that n_{MAX} is large but not "infinity" as the neutrino mass is "small" but not zero.

From dimensional (unit) analysis, the basis of quantum mechanics becomes mass x volume-of-space.

We can see that the ground state of Hydrogen = -13.6eV and is exactly $E_B / 50$ (mass-boundary) on the planet surface.

Then for the Hydrogen ground state,

$n = \kappa c + p(n_{MAX} - \kappa c)$ where $p = p_{unitMV} = 0.02$ for Hydrogen. Then

$n \approx pn_{MAX}$.

We can write a general expression for Hydrogen:

$E_{mb} = E_B - pE_B / n^2$ or

$E_{mb} = E_B (1 - p / n^2)$

For the more general atomic case of massive elements having $p > 1$, we can write the expression for ground-state energy:

E_{GND} (energy) $= E_B (1 - p) = E_B$ (energy / mass-vol) x (1 - p) (mass-vol) where a negative value for E_{GND} implies greater-than unit mass-volume.

Separately, we can derive a universal expression with help from dimensional (unit) analysis.

For **any square law** force:

$F = K_0 k (l_1, l_2) r^{-2} = K'_F (nb)^{-2}$ and $B = b^{-1} = sec^2 = a^{-1} = $ Mass / Force, then

$1 / b = $ mass $(1 / K'_F) (nb)^2$ or we can write

$K'_F = $ mass x n^2 x b^3

And better stated:

Mass x Boundary Volume $= K_F / n^2$

And this defines the allowed ratios E / E_B.

(As a straight-forward example, the square law mass-force: $K_F = 1$ x $K'_F = G m_1 m_2$.)

We assert the physical relevance as a mass-volume traversal through one spatial boundary $b^3 = (1.111E\text{-}17)^3$ met^3 for a combined mass bound in space by a square law force acting at radius $r = n_R b$.

We interpret the chemical relevance as a molecule having lower E / E_B (a lower mass x volume) than the sum of its constituent atoms.

In the absence of time t, the real three-dimensional combined forces on a single particle with mass m and charge q become the summation of square law forces:

$$F_{total} = \Sigma \, F(n_R \times b)^{-2}$$

summing over mutually allowed n_R for masses m having charges q within a single event 1xB ~ event meter^{-1} = 1 / b ~ meter event^{-1}.

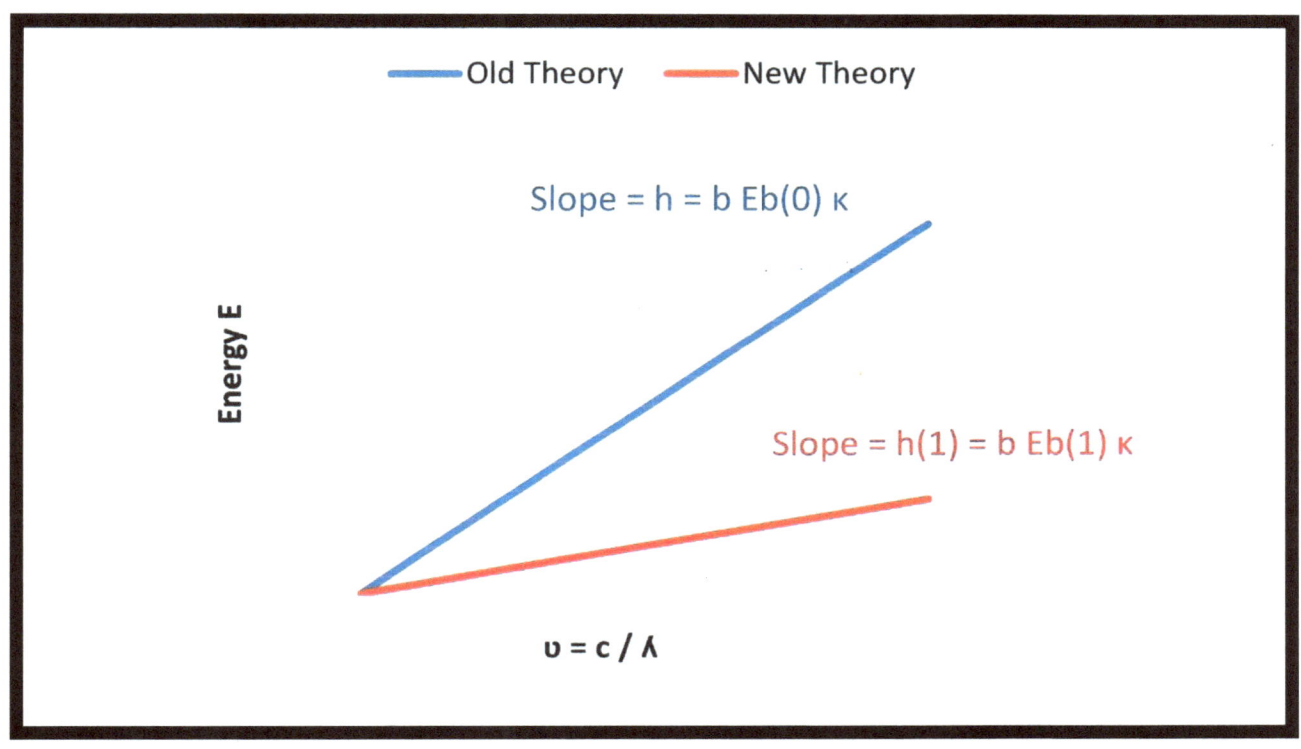

Examples of <u>Available Sacrificial Bonds</u> are the C-H and N-H bonds in the protein binding shown below. State changes can lead to the exact quantum barrier energy E_B but should require an independent transition in order to achieve the exact quantum sum.

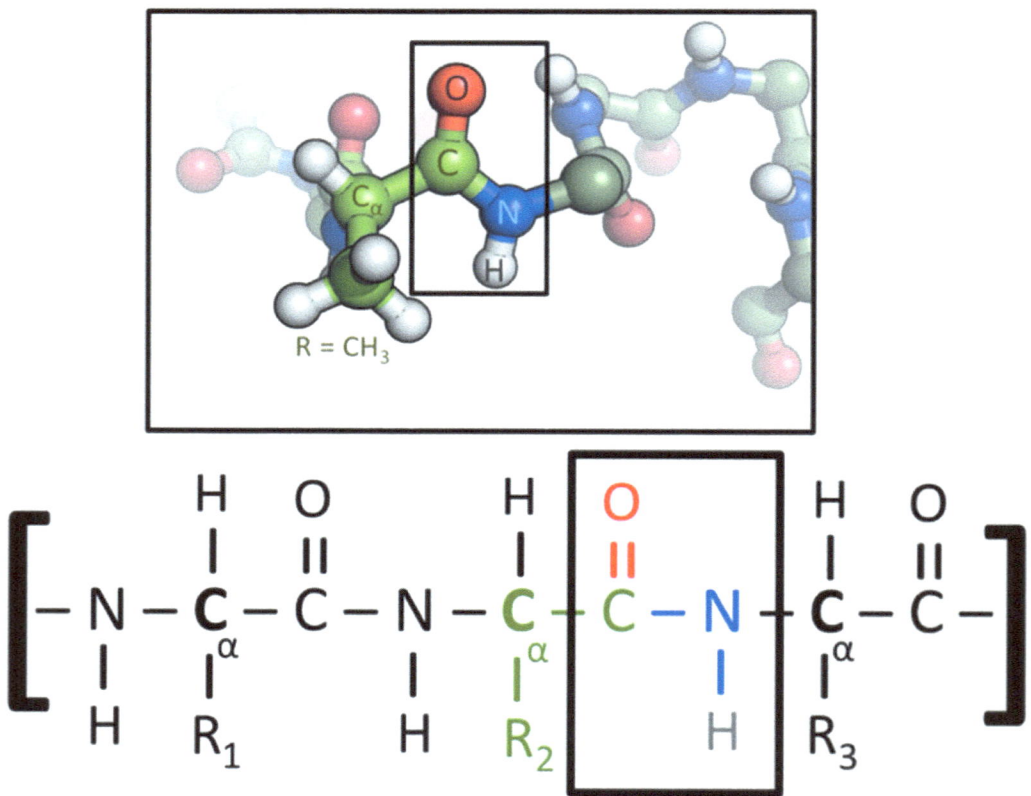

One-Dimensional Electron Example

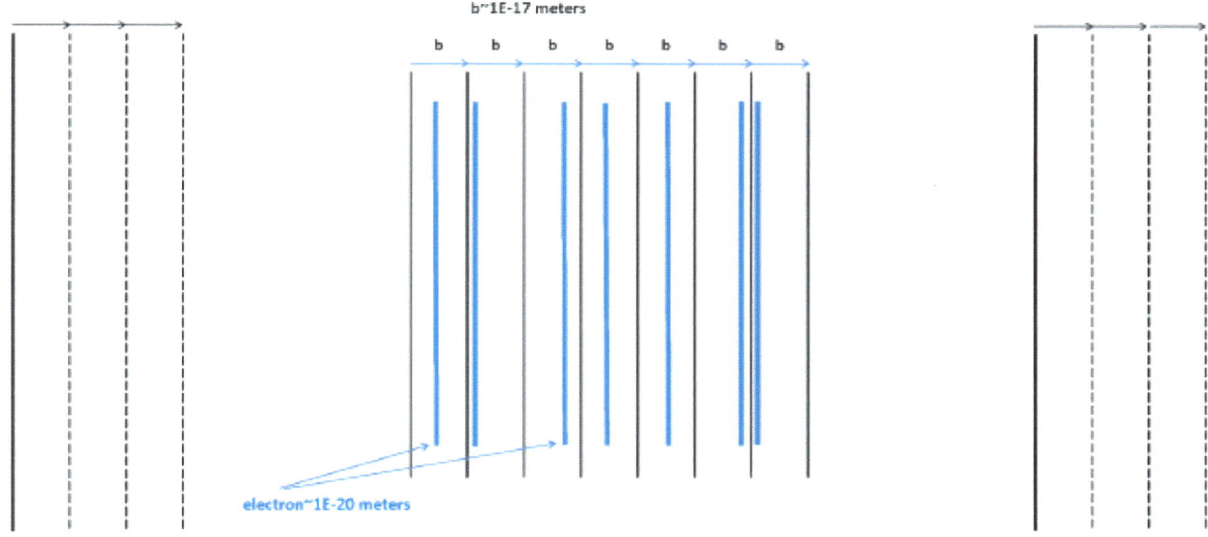

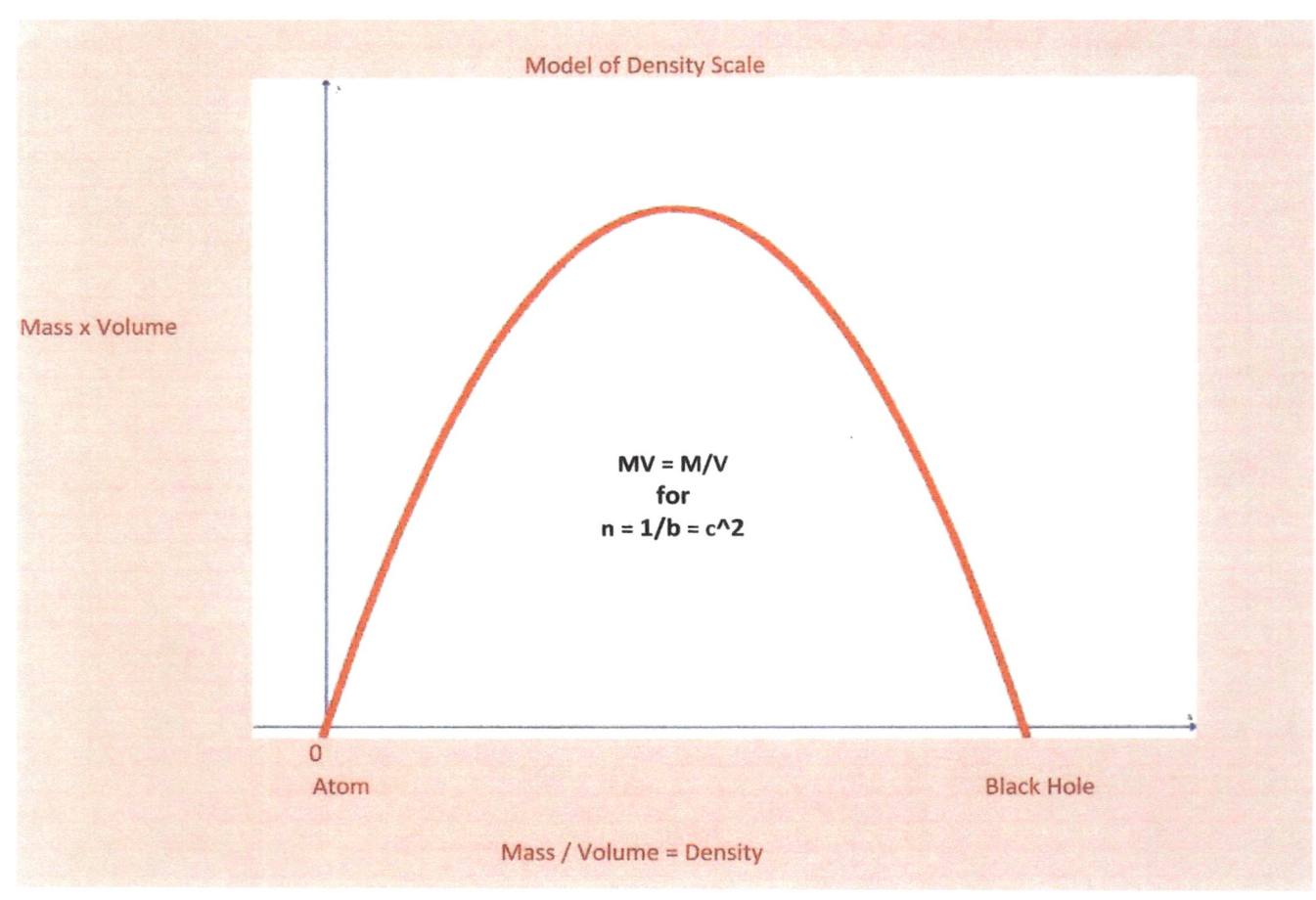

The Expanded Theory

IV. The Expanded Theory

In the absence of time t, we should review spatial intersections given the establishments of appendices A through C and also D through E.

Appendix G

Boundary Conditions and Density for Black Holes

As the intersection of one dimensional space (a line) with two dimensional space (a surface) is a single point with zero dimension, the intersection of two and three dimensional space is a line with one dimension, and the intersection of three and five dimensional space should be a surface with two dimensions, then the intersection between five and eight dimensional space should be three-dimensional (observable in 3-dimensions) and is suggested by the spherical volume of a black hole.

From Appendix E, the maximum allowed energy-event is 2.700E+25 J.

Then E-sub-B at a black hole surface should be bounded by the maximum allowed c^3 J kg^{-1}.

For the hole surface:

$E_{Bmax} = Gm_H r_H / r_H{}^2$ or E-sub-Bmax = G x m / r for the hole, and

c^3 = G x m / r relating to the hole, or we can write

$$m_H / r_H \leq c^3 / G$$

or

$$m_H \leq r_H c^3 / G$$

where G = 6.673E-11 met^3 kg^{-1} sec^{-2},

r-sub-H has units meters, and

c (3E+8 numerical) has units J$^{1/3}$.

Then $c^3 / G \leq$ 2.700E+25 / 6.673E-11, and

$m_H / r_H \leq$ 4.046E+35 kg met^-1

for any black hole.

If we let E-sub-B = $\Delta\lambda$ E-sub-Bmax = $\Delta\lambda$ c^3 where 0 < $\Delta\lambda \leq$ 1, and

$\Delta\lambda$ = nb where 1 $\leq$ n $\leq$ 1/b = B, then

$m_H = r_H \Delta\lambda$ c^3 / G or

$$m_H = K_G \, r_H \text{ where } K_G = \Delta\lambda \text{ c}^3 / G.$$

Appendix H

Definition of the "Event Horizon" for Black Holes

Per Appendix G,

$m_H / r_H = K_G$ or m-sub-H = K-sub-G x r-sub-H

where $K_G = \Delta\lambda$ c^3 / G.

The following <u>surface density</u> boundary conditions should apply for any black hole:

1. The hole mass m-sub-H / ("surface-area") of the 3-dimensional <u>2-D surface</u> = m-sub-H / ("surface-area") of the 5-dimensional <u>3-D surface</u> (volume.)

2. Similarly, m_H / (4/3 x π x $r_{H^\wedge}$3) = C_R x r_H^5 where C-sub-R defines the 5-dimensional "surface-area" for 8-dimensional space.

Then boundary conditions require:

$4\pi r_H$^2 = $C_{R3\text{-}8}$ x r_H^5 and 4/3 πr_H^3 = $C_{R5\text{-}8}$ x r_H^5

where $C_{R\,n\text{-}m}$ = C-sub-R for dimension n curving through dimension m and r-sub-H = $r_H = C_{R5-8} / C_{R3-8}$ from boundary conditions.

Then Δλ = C-sub-R3-8 / C-sub-R5-8 or

$$\Delta\lambda = C_{R3} / C_{R5}$$

where C-sub-R3 and C-sub-R5 represent the curvature rates of 3 and 5 dimensional space respectively through 8-dimensional space, where the ratio m_H / r_H is proportional to Δλ, and where we assume C-sub-R3 ≤ C-sub-R5.

Then there is only an effective zero-density "black hole" for $C_{R3} = 0$ while the highest density black hole occurs where $C_{R3} = C_{R5}$.

Then Δλ represents the ratio of curvatures of 3 and 5 dimensional space through 8 dimensions for the spatial intersections known as black holes.

The higher the mass density in a spatial location, the more the effective radius of curvature should change. With dense enough matter, then curvatures among dimensions become more closely equivalent as density becomes large.

To visualize in two dimensions, πr^2 and 4πr^2 are both two dimensional surfaces that curve in 3-dimensions. The curvature (lack of) for a flat circle is 0 while the curvature for the closed spherical surface is 1.

If we assume C-sub-R5 is closed (curvature 1) in 8-dimensions, then C-sub-R3 has the possible range 0 → 1 in 8-dimensions where 0 represents no intersection at all and 1 represents a closure of the five and eight dimensional intersection.

To see/observe the intersection (black hole) it must be at least a 5 and 8 dimensional intersection (3-dimensional) or constitute the five dimensional surface intersection represented by the integral ∫4πr^2dr throughout r for the continual surfaces.

Then the "smallest" black hole is the "least dense m_H / r_H" black hole having Δλ ~ 0 but still large enough to represent an intersection of 5 and 8 dimensional space.

Then the boundary condition is a single event:

$C_{R3}min = 1 / cB$ where $B = 1$ and

$\Delta\lambda min = 1$ meter / c meters $= 3.333E-9$.

Then

$c^2 / G \leq m_H / r_H \leq c^3 / G$ or

$c\wedge2 \leq m_H / r_H \leq c\wedge3$ kg met$\wedge$-1

or we can write the expression in the Schwarzschild[vi] form,

$r_H = k(\lambda) m_H G / c\wedge2$ meters

where $k(\lambda) = 1 / c\Delta\lambda$.

Appendix K

Spatial Curvature

Appendix H defines the mass-radius relationship as observed in three-dimensional space:

$$m_H / r_H = K_G(\lambda)$$

where $\lambda = \Delta\lambda = C_{R3} / C_{R5}$

and represents the ratio of curvatures from 3-dimensional space and 5-dimensional space through 8-dimensional space respectively.

We assume, for the three dimensional intersections, that $C_{R5} = 1$.

The minimum $C_{R3} = 1 / c$ and the maximum $C_{R3} = c / c = 1$.

Allowed quantum are then n / c for $n = 1$ to c.

The minimum (least dense) intersection is an intersection among 3, 5 and 8 dimensional space where $C_{R5} = 1$ and $C_{R3} = C_{R3}(min) = 1 / c$.

The next "largest" (more dense) intersection should occur for $C_{R3} = 2 / c$ and so on.

The most dense intersection occurs where $C_{R3} = c / c = 1$ and represents a closed third dimension in both eight dimensional and five dimensional space.

To visualize curvatures, the diameter of a circle $= d$ is a straight line with curvature

$C_{R1-3} = 0$ while the circumference (length πd) closes upon itself (runs into the back of itself) and has the curvature $C_{R1-3} = 1$.

The curvature C_{R2-3} is closed in 3-dimensions visualized as a spherical (or elliptical, not reviewed in this scope) surface area that has closed itself around a center-of-mass c_M.

The two dimensional surface does not alter or "grow" in three dimensions, but the one dimensional line, e.g. the straight path of a distant comet or ray of light ($C_{R1-3} = 0$) or the line of a planetary satellite ($C_{R1-3} = 1$) both curve (or bend) around mass in three dimensions to the two extreme degrees of curvature.

Then the ratio $m_H / r_H = K_G$ should represent a curvature of three-dimensional space through eight-dimensional space.

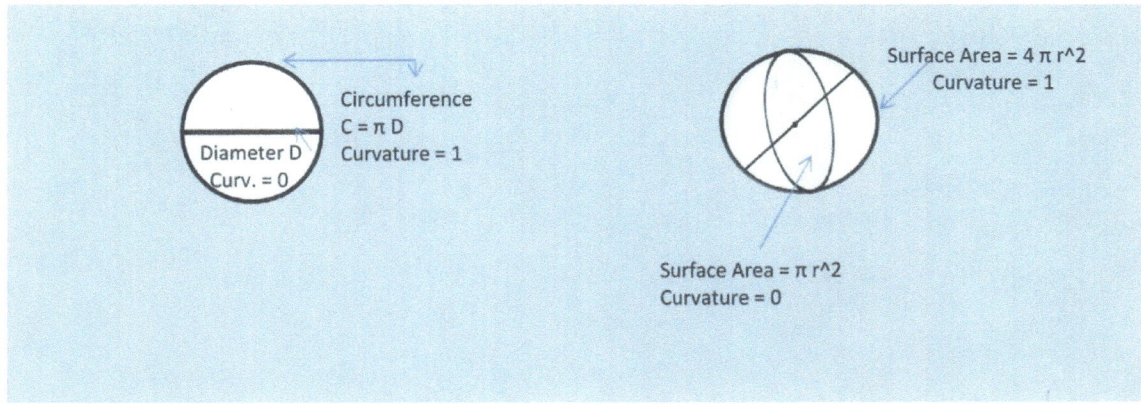

Appendix L

Mass and Dimensional Geometry

From the equation $E_B = a_G$ J kg^-1 b_n^-n x b_n (= 1.089E-16 on Earth surface,)

and from the definition of physical events in dimension n = B where B ~ sec^2 in 3-dimensions and sec^(D-1) per appendix C, then b_n has the following value:

3-dimensions: b_3 = 1.111E-17 meters (per the main text)

5-dimensions: b_5 = b_3 / c = 3.703E-26 meters

8-dimensions: b_8 = b_5 / c^2 = 4.114E-43 meters and so on.

Per Appendices G, H and K, the black hole geometry is a function of mass and becomes a series of symmetrically-closed concentric surfaces having internal densities:

8-dimensional volume in 3-dimensions = 4 / 3 π r_8^3

5-dimensional volume (= ∫ (r_{5-8}-to-r_5) 4πr^2) = 4 / 3 π r_5^3 (including the volume of 8)

3-dimensional volume = 4 / 3 π r_H^3 where r_H = r_5 (including volumes of 5 and 8)

and

the boundary condition for the most-dense black hole is then:

m_H / (4πr_H^2) = m_H / (4/3 πr_H^3)

where r_H = m_H / K_G, then

m_H(max) = 3K_G kg and the corresponding

r_H = 3 km.

The black hole mass m_H for the general case r_H = r_5:

m_H(λ) = ∫ (from r_5 – r_8 -to- r_5) 4πr^2 dr = 3K_G(λ)

where

$\lambda = r8 / r5,$

$r_H = r_5$ and

r_8 is the 3-dimensional-radius of zero-mass 8-dimensional space ($b = b_8$) at the center of the hole.

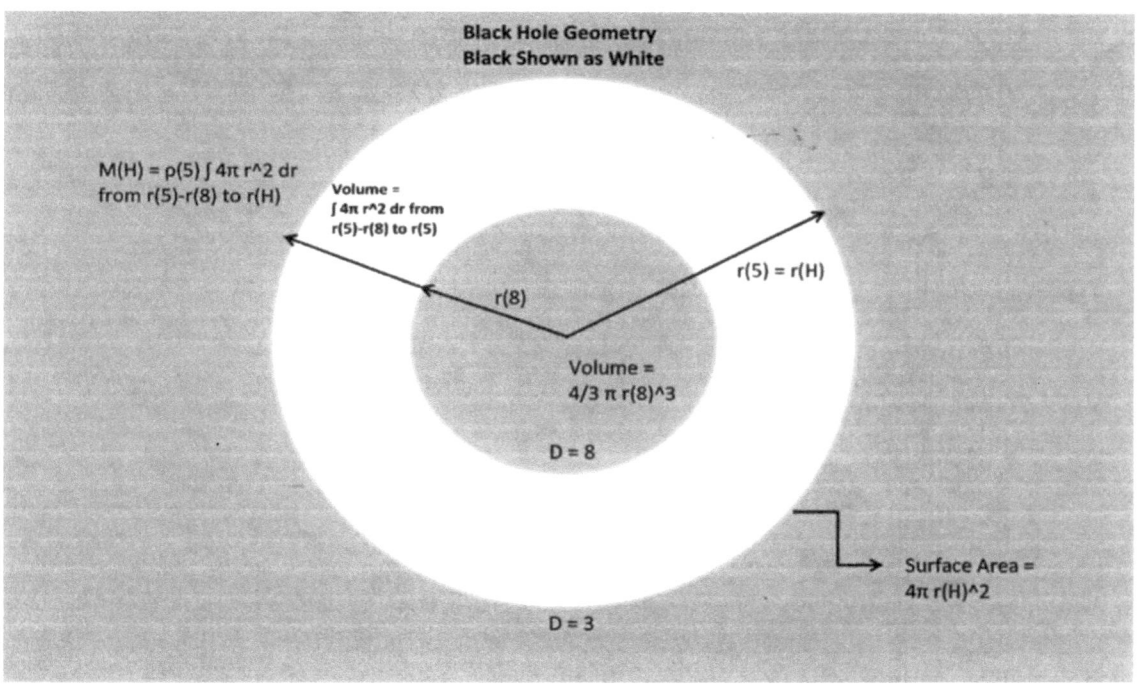

Appendix M

General Spatial Intersections

Per Appendix L, the adjacent intersections to three-dimensional space D = 3 are with dimensions D = 2 and D = 5, and the intersection between three and five dimensional space is a two-dimensional surface, e.g. a closed surface around a large mass M.

There are infinite concentric closed surfaces around the mass M, so we can define a surface "intensity" or transmission-ability proportional to r^-2:

$$T_s = G' F_G(r)$$

or more generally,

$T_s = \Sigma\, G' F_G(r)$ for masses M at distances r.

In the absence of continuous time t, then the 3-dimensional spatial sequence progresses through 5-dimensional space. In order to access a prior 3-dimensional spatial frame, an interaction is required through a two-dimensional intersection.

As an example, the reader can think of any important memory. Subsequently, a photograph-like image immediately appears to the reader's mind. In the absence of continuous time, the three-dimensional spatial frame would need to be accessed in order to review the memory.

The access required is through a two-dimensional intersection and through five-dimensional space in order to obtain information contained in a prior three-dimensional spatial frame. The previous spatial frame(s) is still there. It didn't go anywhere. The frames should be available for continual access.

There can be access/communication at the speed of light c by energy state transitions, including hyperfine transitions, in hydrogen atoms and other molecular state transitions as well as low energy magnetic dipole moment interactions from molecular charges within large hydrocarbon (organic) molecules.

Arguably, not all memories are self-contained. Electro-magnetic communication can take place as a result of hydro-carbon molecular energy state transitions and with the surface(s) having the combined intensity $\sum T_S$ as above. The communication rate (not the perceptibility rate) should be the speed of light c.

$\sum T_S$ is better defined as follows for an altitude A above the surface of mass M:

$$T_S = (r_{CM} \rightarrow r_S) \int G'(\varepsilon, r) \, F_G(r) \, dr + (r_S \rightarrow r_A) \int G'(\varepsilon, r) \, F_G(r) \, dr$$

$$+ (r_A \rightarrow \text{infinity}) \int G'(\varepsilon, r) \, F_G(r) \, dr.$$

Appendix N

Nuclear Forces

Consider two possibly adjacent protons interacting through two 3-dimensional square law forces, the charge force and the mass (gravity) force. The forces act along straight lines (1- dimension) through square laws (2-dimensions) in 3-dimensional space.

For the gravity force, the magnitude is $G(1\ amu)^2 / r^2$ where $r \sim b$.

Per appendix E, then mass x volume $= K_F / n2 \sim (1E\text{-}11)(1\ amu)^2 \times 4/3\ \pi\ (3/2\ b)^3 / n2$ or

Mass x volume $\sim (1E\text{-}11)(1E\text{-}27)^2(1E\text{-}17)^3 / n^2 \ll 1E\text{-}85$ for any n

where we have used the three-dimensional special relativity law mass $_{MIN} = E_B / c2$ to calculate the minimum bound on mass x volume $\sim 1E\text{-}85$ in a Joule system of measurement.

Therefore, the weaker of the two square law forces, by itself, breaks the allowed three-dimensional mass-volume rules for the extent of any nuclear size.

We suggest the atomic nucleus to be higher dimensional (b < 1.111E-17 meters per appendix L) having an exterior 2-dimensional-shell intersection with 3-dimensional space (volume.)

In that case, nuclear forces should not be bounded by one-dimensional square-laws; instead, they would be minimally bounded by two-dimensional cube-laws and should be relatively large.

The Natural Theory
The Non-Technical Version

V. The Natural Theory: A Non-Technical Version

Our concept of continuous time is deeply rooted in our perceptions. In this segment, we consider a case where time is not in fact fully continuous and where time can be equally represented as a contiguous spatial "growth."

How Could Space Move, Progress, or Sequentially Grow?

We start with the Fibonacci natural rate of growth. The Fibonacci spatial sequence is best illustrated by photographs and diagrams:

A photograph of an aloe plant as it has naturally grown:

A photograph of a seashell:

A diagram of the Fibonacci spiral with curvature ρ, as it mathematically and exactly matches the natural photographs:

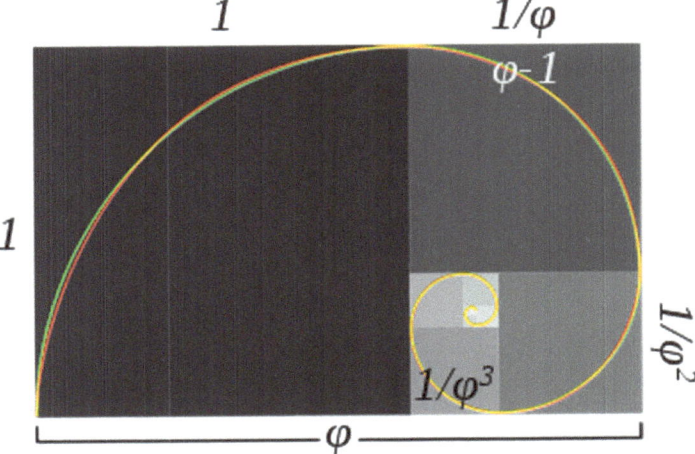

The mathematical match to nature does not stop on the planet surface:

See Appendix Supplement I for a more intuitive and mathematical treatment of the Fibonacci infinite spatial sequence.

An aside: In the year 1202, Leonardo Fibonacci was an Italian mathematician from a wealthy family. He discovered the work of an ancient (B.C.) Indian mathematician who had postulated the future population of rabbits given certain starting conditions.

We have been viewing two dimensional surfaces (photographs) from our personal three-dimensional perspective. We have seen a one dimensional line curve in two dimensions on a page of paper, on a chalkboard, or on our viewing display surfaces. We can also imagine the three dimensional result, for example the aloe plant growing nearby for us to see every day.

A Closer View of the Spiral:

We return our attention to the diagram of the geometric spiral (on page 43) having the symbol ρ in its measurements. (ρ is simply nomenclature for the ratio 0.618 defined in Appendix Supplement I.)

Since we live in three dimensions, we can easily see the two dimensional intersection within the Euclidean spiral, i.e. the linearity of the spiral intersects with three adjacent two dimensional regions per the diagram.

Space itself, as we know it, is three dimensional. If we lived in five dimensions, we could easily see the three dimensional intersection of a two-dimensional "spiral" with "five" regions of three dimensional space per the natural sequence.

We do not live in five dimensions; instead, we live in three and we cannot fully "imagine" a higher dimensional state.

An aside: If we lived in the year 1492, we may have believed the Earth was flat. We would have the same mindset as two dimensional creatures. We would have no idea that the Earth surface in fact was curved in the next highest spatial dimension (D = 3) until Columbus and others discovered they could sail ships back to where they started by never returning; instead, simply continuing on their journey brought them back to where they came from.

 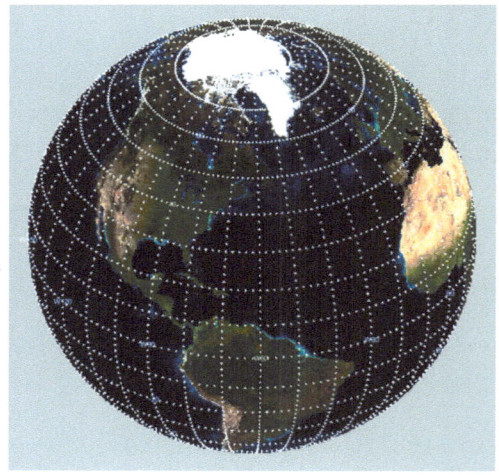

The Renaissance mariners confirmed the Earth surface was two-dimensional, but also that the surface "wrapped-around" itself in a third dimension (it is bent or curved back onto itself in three dimensional space.)

The two dimensional surface is in fact "closed" upon itself through the next higher dimension.

Time vs. Sequential Spatial Frames or Sequence of Events:

Time is definitely a great and useful concept and is a valuable measurement tool that tells us about the relative spinning position of the Earth with respect to the Sun (day vs. night) and also the relative position of the Earth's precession (slight wobble) on it's orbit around the Sun (seasons of the year.) Time defines clocks and the calendar. Time is embedded in our language(s) in terms of past, present and future tenses, expressions regarding sequences, etc. Time is embedded in our perceptions.

But time is not required to exactly match our perceptions except in the macroscopic sense. In particular, time is not required to be a "real and continuous" variable. The continuous concept of time can be replaced by a slightly contiguous view of spatial growth with the same physical result. A sequential motion of spatial frames is a new concept but yields the same macroscopic every-day result.

Anyone Could Tell the Difference?

We would never know the difference.

We have evolved on the Earth's surface over 700 million years. Everything we are and everything we think is based on our experience and evolution here. No one has the right to assume everything in the Universe needs to be the same as our Earth-bound perceptions and concepts.

To view a plot (graph) of our neurological synaptic speed, please see Appendix Supplement II. The horizontal axis units are in milli-seconds.

We do not have fast enough synapse "switching speed" to discern between sequential space (per the main text) and the concept of continuous time. Similarly, we also perceive a movie (motion picture) to be a continuous motion even though we know it is only a contiguous series of still photographs placed in the right spatial order or sequence viewed quickly enough to bypass our perceptions.

Space is real. We have evolved in and through space. Mathematics comes from our surroundings, not the other way around.

While we can mathematically achieve 2 from 0 and 1, we cannot physically achieve two from nothing and something. The adjacentcy of 0 and 1 can physically produce only one plus one = two. Then two plus one can be subsequently achieved and so on.

Each space quantum should "experience" only each of its boundaries.

The juxtaposition of space is physical reality. The sense of time serves to approximate

physical laws and works well within the bounds of our senses.

Implications of Quantum Barrier Energy:

Aging Without Time

Except as a measurement tool, there is no real physical time t; instead, time is a name for energy per sequential spatial frame.

We live (have evolved) on the planet Earth surface through our chemical composition of hydro-carbon (organic) molecules consisting of hydrogen/water and carbon/ash. We should grow and subsequently die due to the quantum energy requirement at each spatial barrier. Arguably, energy state transitions in hydrocarbon molecules are the most likely vehicles for the aging process.

For a simplified proof of hydrocarbon energy states, please see Appendix Supplement III.

The barrier energy is a quantum energy meaning the energy (including energy from us) cannot be lost unless it is an exact quantum state or multiple per the main text. An energy that is slightly different from a quantum state means 0 = zero energy loss. The energy (and aging) loss must be quantum (parced into finite pieces) as opposed to a continuous (could have any value) energy event.

Energy loss should effectively slow down or even stop for our physiology in the case we were located in a region of very low gravitational force. Of course, no one wants to live in a space station or on the Moon.

The Earth-surface quantum energy sum could theoretically be altered so that little or no energy was lost through the spatial progression.

Energy production

Energy production or accumulation should (arguably) be free-of-charge in the proper spatial locations of low gravitational force, for example in the region where the Moon and Earth gravity offset one another to create a very low gravitational force.

We can think about an energy transforming turbine at the Hoover Dam. With less or no E_B-energy to stop the turbine from spinning, the turbine should not need the waterfall in order to turn.

Interaction, Access and Communication

With no actual physical time t, then everything progresses through the natural spatial sequence of physical events.

For example, memories may not be stored captively inside the mind; instead, the prior spatial frames (memories) are accessed through the communication ability of the brain's hydrocarbon energy states that communicate through spatial dimensions using low electro-magnetic energies. Please see Appendix Supplement IV for a technical description that is less mathematical than the main text.

De Ja Vu and ESP:

These things are not "weird"; instead, they are results of spatial intersections in the absence of time t.

For example, when you feel as though "I've been here before!" the reason you feel that way is likely because you have in fact been in that same spatial intersection "before."

Supplemental Appendix I. Our Natural Growth

The Fibonacci sequence is the sequence of numbers (integers) as follows:

0,1: 1, 2, 3, 5, 8, 13, 21, 34, …

and in fact represents, among other things, a rabbit population given certain starting rules, etc.

where the seed values (initial numbers) are 0 and 1 as shown,

and then each subsequent number is the sum of the prior two numbers.

For example (for n = 3, meaning the third sequential number) 1 = 0 + 1.

Next, 2 = 1 + 1.

Then 3 = 2 + 1.

5 = 3 + 2, 8 = 5 + 3, and so on.

The sequence represents a start (nothing to something) followed by a growth that grows only from itself and the immediately preceding value, e.g. 3 grows from the juxta-position of itself and the only other position it is adjacent to, i.e. 2. In that case, the next sum is 3 + 2 = 5 and so on.

The ratio, as the numbers in the sequence become large, of the given sequential number divided by the most previous sequential number is $\rho = 0.618\ldots$ as approximated by 5 / 3, 8 / 5, 13 / 8, and so on.

This represents the sequence of nature, i.e. the sequence of natural growth.

Supplemental Appendix II. Our Synapse Limitation

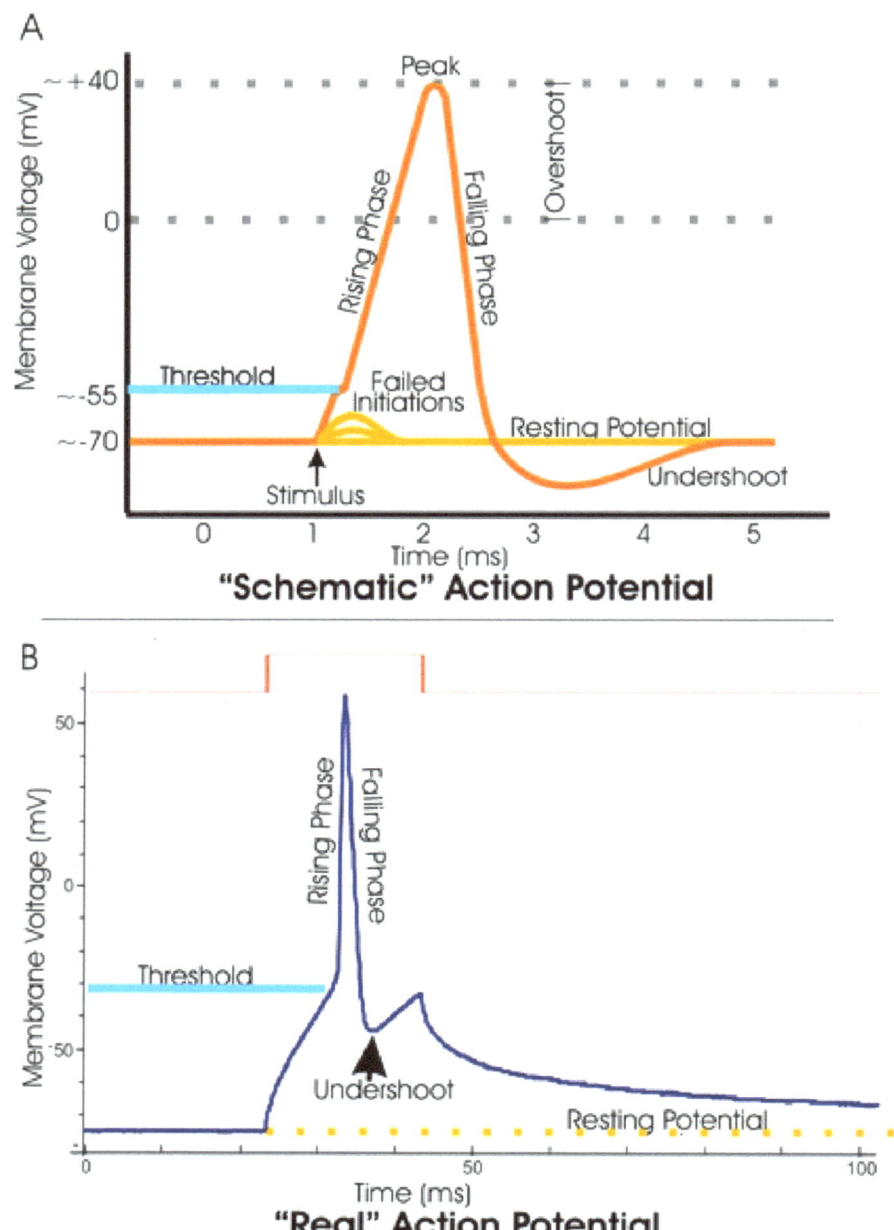

Supplemental Appendix III. Our Chemical Nature

Per the main text and Appendices D and E, the physical nature of atoms and molecules including hydrocarbon molecules and their proportional energy states (quantum mechanics) is related to mass and gravitational force within space.

Earth surface E_B (sequential boundary energy per unit mass) replaces the concept of continuous time. The energy E_B we experience is an exact quantum multiple of the known Hydrogen atomic energy states. In that case, the hydrocarbon molecular energies are susceptible to deteriorating through the quantum requirement per spatial frame.

Then life as we know it should grow and die through quantum energy states of hydrocarbon organic molecules.

Hydrocarbon death should be accelerated by a larger gravitational force (closer proximity to higher mass) while hydrocarbon extended life should be a result of less mass and a lower gravitational environment.

Common quantum energy transition sums within hydrogen alone are $13.6eV = 10.2eV + 3.4eV$ and are an exact multiple of the 680eV quantum energy requirement.

Then our evolution is a result of the Nature of Space.

Supplemental Appendix IV. Our Communication Ability

Per Appendix M in the main text:

The adjacent intersections to three-dimensional space $D = 3$ are with dimensions $D = 2$ and $D = 5$, and the intersection between three and five dimensional space is a two-dimensional surface, e.g. a closed surface around a large mass M.

The best example is the surface of the planet Earth. There is no need to consider exact altitude A from the spherical surface.

There are infinite concentric closed surfaces around the mass M, so we can define a surface "intensity" or transmission-ability proportional to r^{-2}:

Then the transmission ability T_S is proportional to $1 / r^2$ similar to the physical laws of sound, i.e. the farther away from the source you are, the less you can hear, etc.

In the absence of continuous time t, then the 3-dimensional spatial sequence progresses through 5-dimensional space. In order to access a prior 3-dimensional spatial frame, an interaction is required through a two-dimensional intersection.

As an example, the reader can think of any important memory. Subsequently, a photograph-like image immediately appears to the reader's mind. In the absence of continuous time, the three-dimensional spatial frame would need to be accessed in order to review the memory.

The access required is through a two-dimensional intersection and through five-dimensional space in order to obtain information contained in a prior three-dimensional spatial frame. The previous spatial frame(s) is still there. It didn't go anywhere. The frames should be available for continual access.

There can be access/communication at the speed of light c by energy state transitions, including hyperfine transitions, in hydrogen atoms and other molecular state transitions as well as low energy magnetic dipole moment interactions from molecular charges within large hydrocarbon (organic) molecules.

Arguably, not all memories are self-contained. Electro-magnetic communication can take place as a result of hydro-carbon molecular energy state transitions and with the surface(s) having the combined intensity T_s as above. The communication rate (not the perceptibility rate) should be the speed of light c.

References

i	Erwin Schrödinger, Quantisierung als Eigenwert Problem, 1926.
ii	Werner Heisenberg, Über die Grundprinzipien der 'Quantenmechanik, 1927.
iii	Albert Einstein, Zur Elektrodynamik bewegter Körper, 1905.
iv	Leonardo Fibonacci, Liber Abbaci, 1202.
v	Max Planck, The Genesis and Present State of Development of the Quantum Theory, 1920.
vi	Karl Schwarzschild, Uber das Gravitationsfeld eines Massenpunktes nach der Einsteinschen Theorie, 1916.